BIO-CLIMACTIC

VERTICAL-HOMELESS-HOUSING-FARM

PROJECT:

BY

Thomas Charles Bass, M.S.M.

MASTERS OF SCIENCE IN REAL ESTATE MANAGEMENT

"A California Real Estate Broker since 1992"

https://www.linkedin.com/in/thomasbassrealty

(800) 604-8436

DRE #: 00961512

Table of Contents

"The Vision"
Bio-Climactic
Vertical-Homeless Housing-Farm

Sea-Water Vertical Farm, 2014
"We must not perish for a lack of vision… Proverbs 29:18"

I NEED SOME HELP Y'ALL!
Click Here & Listen to Teddy Better Explain This Page!

- This proposal to develop vertical homeless housing complexes is for you "the people." If you agree with the plan than buy the e-book again. Also, we get paid every time you read this book on Kindle Unlimited. So, go to our  music therapy page and listen, listen, and read… read… READ!

- Every time you buy this E-book, 50% of the net proceeds goes toward the support, research, & development of vertical homeless housing.

- TOGETHER "we can change the world… just you & me…" Access the music page on the next page of this e-book from Kindle Unlimited as many times a week as possible!

- "We can't do it alone." All "you" have to do is "Wake up! Activate! "The world won't get any better if we just let it be… We can change this… you & me." A big project like this means we need big help from you!

- Help us grow our network by requesting to connect with us at:

  https://www.linkedin.com/in/thomasbassrealty

"The single greatest lesson the garden teaches is that our relationship to the planet need not be zero-sum, and that as long as the sun still shines and people still can plan and plant, think and do, we can, if we bother to try, find ways to provide for ourselves without diminishing the world."
Michael Pollan, The Omnivore's Dilemma: A Natural History of Four Meals..."

Music Therapy for the Homeless & You!

*Take Your Time *Deja Vu	Sweet Love *The Closer I Get to You	*Heaven Must Be *Unforgettable	California Dreamin'
*Love Calls *New Love *We Both Deserve Each Other's Love	I Made it Through the Rain I'll Always Love You	*Stand *Better Days *Amazing Grace *We Fall Down	I'd Really Love to See You Your Smile
*Better Days *We've Only Just Begun	*Feel Like Going On *Dreamin' *Come, Let Us Go Back...	*Time to Come Back Home *Early Morning Love	If This World Were Mine
*Wishing On a Star *We're Not Gonna' Take It!	*As *If You Believe *Sadie	*Love Ballad *Inside My Love *I Feel Like Going On	*Lord Give me a Sign. *We Are Soldiers
Over the Rainbow Stay With Me	*You are My Lady *Cruisin'	* Summer Breeze *I Apologize	In the Air Tonight

Come back to this page daily to stream music proven to improve the mental health of stressed-out human beings.

➤ I have now enlisted you as a homeless-music-therapist! Resistance is futile because your spirit, like mines, knows when to help! This musical matrix is a key in our strategic plan.

➤ Here's how! Listen carefully to the vibe of each and every songs. Now it is time for you to add your favorites by sending email requests to butterflyrealty@aol.com for the "musical expansion of our matrix society."

➤ We will continually add all musical genres where the vibe and message fits. But, expect to hear the sound of crickets if you email a request for a Twisted Sister melody as you are not on the same page as the rest of us.

- Oops, and exception to that rule would be "We're Not Gonna' Take It!" And, we will go way-out into all genres to expand our musical society, like, adding DMX's Lord Give Me a Sign, or Elvis's Amazing Grace. ☺

- Religion is not a requirement, but your message must be on point with the matrix's vibe in order for us to accomplish our goal of building vertical homeless housing.

- If we believe together, we will grow our musical-matrix into a revenue producer just by listening, and believing together! It's a spiritual fact…just listen to this link to get a better understanding! And, as our musical matrices grows, so multiplies our vision along with our vertical homeless housing project!

- You can immediately use this E-Book to help the homeless feel better by turning on this healing music near them! Go through our music pages often, listening to each song, to help us!

- Be sure to open our e-book from your Amazon Kindle to access the music each time you visit. Please note expansions to the matrix will be added at the end of this book.

- The mental well-being of people who are homeless or under tension will improve by taking heed to the music therapy on this page, as borne out by data from the American Psychiatric Association.

- Many of the homeless do not have access to this music… you can stream music from this e-book via your cell phone with them the next time you eat in a park or are near them… knowing you are helping link them to a better sense of well-being! The NIH & DVA both say music therapy helps the homeless move from isolation to community Lamb, Kass, 1992), (Tennessee Medical Journal, 2012).

ABSTRACT

Research surveys show that on any given night approximately 407,966 individuals are homeless across the United States, and live in transitional housing, or shelters, according to a, 2010, United States Department of Housing and Urban Development Study (Pauette, 2011). This HUD report also disclosed that almost 50% of the homeless have co-occurring substance misuse problems, while nearly 30 percent of them are chronically homeless, and suffer from untreated mental health illnesses. (Pauette, 2011).

The purpose of this presentation is to identify data driven interventions, and strategies that more efficiently address the uncared for needs of the homeless. To boot, the following research highlights some of the heavy costs government agencies incur as a consequence of ineffective homeless interventions. This research shows citizens and their governments a more efficient way to spend the current costs of approximately $40,000 per year allocated for homeless housing (Moorhead, 2012).

Therefore, if an example group of 400,000 chronically homeless families currently cost the federal government upwards of approximately $16 billion yearly, and $160 billion every decade, to provide housing in a non-cost, efficient manner, then it would be more efficient to invest $7 billion in a vertical-homeless-housing project. This project improves the inhabitant's outcomes as demonstrated by this research, and it rewards the $7 billion or higher investor up to a 20% equity share of the high-profile building, and 20% of the overall yearly revenue of the project.

. The self-sustained project's Dragonfly building will be designed by Vincent Callebaut, and span 132 stories high (Cilento, 2009). This, one city block square, vertical farming homeless center, united with a vertical farm will be powered by renewable energy, and include 28 different agricultural fields. It will produce 3,500 tons of fruit per year, along with 140 tons of catfish, tilapia, along with quality chickens, and cattle feeding 35,000 people yearly, as well as providing an indefinite source of self-sustaining revenue (Kain, 2014). The initial cost of this self-sustainable project is thought excessively high by investors to justify due to the high cost of the initial energy required for the agricultural features (English Online, 2015), (Michael, 2014).

But, when the vertical farming is combined with a homeless project, and the subsequent ineffective homeless servicing costs for repeated chronic homelessness replaced, the cost of unnecessary incarceration prevented, and homeless related mental health care costs re-directed into vertical-housing projects, then the innovation becomes economically viable, as supported by this research. The project's derivative return on investment can range from about $4 to $14 billion (Banerjee, 2013, pp. 46-47).

Vertical housing farms are not in jeopardy of severe weather events, and can be installed in rural, urban, ghettos, desert countries, and on space stations, and will house, feed, and train over 10,000 homeless yearly for high demand agricultural jobs (Vertical farming, 2015).

(Huff Post, 2014)

Acknowledgement

When I was a kid, enduring in the black-area of a rural town named "Sweet-Home" on the outskirts of Little Rock, Arkansas, my mother and grandmother grew all of our food in a giant sphere in our backyard. They raised chickens, and several other animals, so that we could always have something to consume.

My family became so efficient at farming that we oftentimes had a surplus to sell to local stores in Little Rock, and surrounding cities. My mother would always make sure to send me door-to-door to help scores of our most needy elderly neighbors with three pre-prepared meals for their everyday food needs.

We were determined not to depend on the government, but to help each other when we got into a position to serve. But, more importantly, my mother would make sure that when I delivered the food, I took the time to respectfully listen to each and every older neighbor's story.

That was a difficult request because old, and lonely folks love to talk, and Lord would they talk, while my cousins would be patiently waiting for me to finish delivering food, so we could kick-off our daily, over-hyped, baseball game that was always considered the ""baseball-finals-of-the-world."

I dutifully made the high-level business decision to finish the task of presenting the food, and taking heed and listening to the elderly neighbors, versus sustaining a possible severe concussion from a country-styled whipping' from my sort-of-muscular, African-American momma… ☺

God has given us all gifts. Mine is helping people learn, and acquire real estate. With my real estate gifts comes a responsibility, and that does not mean simply acquiring money for myself. I've earned money in real estate only to lose all the money, and a great deal more in my personal life, as a consequence of not remembering my work was "NOT" more important than my folk, and my community.

What I have found out from life's on-the-job-training is to first get-back-up off the rug after the knockout punch because "chickens do come back to roost,' according to my grandma.

Nowadays, I have re-ordered my steps to work toward the development of a sustainable, affordable, housing project that is worthy of needy families. And, if I am successful, as I normally am, I will be sure to walk-in on each-and-every family within this project, daily, to find-out how they are getting along, and take the time needed to listen to them, like my "momma" taught me.

HOW TO GET STARTED VERTICAL FARMING

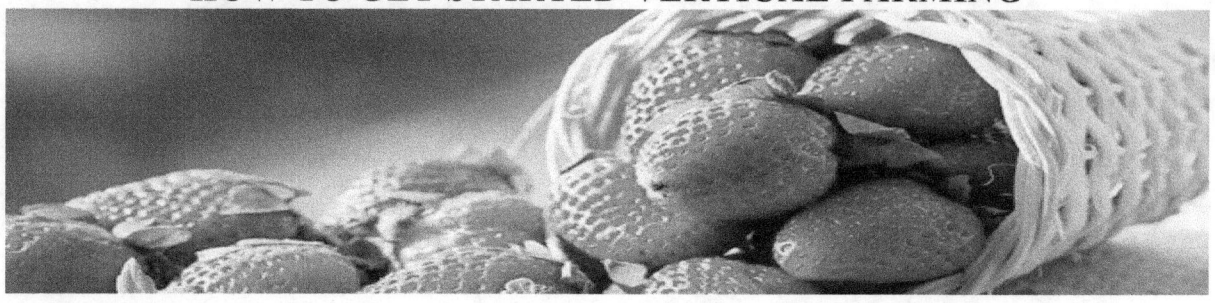

Model 1: ***Home-Based Vertical-Farm*** – Vertical farming is so versatile that it is scalable, and permits you to develop a vertical farm from your domicile with an 8' by 8' space, or as high as a skyscraper. In the event of a catastrophe such as an earthquake, storm, or etc., you can be prepared to sustain your own food supply as a result of developing your home vertical farm.

Architect Ken Yeang says that plant life should be cultivated within confined, mixed-used skyscrapers in order to achieve maximum climate control (Wikipedia Staff, 2015). Also, a mini-vertical-farm you successfully manage at your home can grow up to 30 times the quantity of food compared to a similar sized outdoor farm. An ideal climate must be carefully kept within the farming area to range from 65 to 76 degrees. Most plants perform well with a relative room humidity of about 62 degrees. Nighttime temperatures should run from 55 to 65 degrees.

You can grow rows of agriculture upwards in a vertical manner. Raising your own food efficiently in an uncertain world like the one we live could mean the difference between living and dying if our social system ever breaks down.

Maintaining a climate controlled system for your beginner-farm is key will multiply its productivity, and the quality of food produced within your farm can improve by a factor of 4 to 6 (Wikipedia Staff, 2015). Whatever you grow within your mini-vertical farm could attain a much better food quality than that bought at your local shop if managed properly. Click here for more great information on how to manage and prosper from your own, survivalist, vertical farm.

Model 2: A *conglomerate of research* universities can leverage substantial charitable funding and invest $7 billion to start-up a vertical housing farm. The project can provide further research into the subject while at the same time bringing the school great revenues. School students in varying departments can do research based on the inhabitant's outcomes, and establish exact budgetary norms for agricultural services. Successful revenue outcomes supports any future budgetary needs.

The university immediately owns up to 20 percent of equity in a high-rise building complex, plus the university generates revenue from 28 different agricultural fields producing upwards of 3,500 tons of fruit per year, along with 140 tons of catfish, tilapia, and with quality chickens, and cattle feeding 35,000 people yearly (Kain, 2014)..

Model 3: A group of individual investors input $16 billion for the project. The investors immediately own up to 25 percent of equity in the building complex, plus own up to 25 revenue share. Vertical-housing farm complex grows by developing complex systematically upward into around a 28 farm complex.

Model 4: Government agency investors input approximately $16 billion for project in California. A housing project is then infused into a vertical farm building constructed in California with a size up to 13,068,000 square feet, and an approximate cost of about $16 billion. The investor wins with an equity share of the multi-billion commercial building's value, plus sustains shareholder revenue generated by the vertical farm's agricultural revenue activities. Additionally, the investor wins in taxation of farming revenues, and by sharing in revenue generated from 28 different agricultural fields that will produce approximately 3,500 tons of fruit per year, along with 140 tons of catfish, tilapia, along with quality chickens, and cattle feeding 35,000 people yearly (Kain, 2014).

CHAPTER 1:
BACKGROUND

Cost of Homelessness – Statistics:

Program	Cost Savings per Person	Sourcing Information	City of Study
Eastlake Program	$30,000	(National Alliance to End Homelessness, 2015)	New York
Housing First Program	$31,450	(Carrier, 2015)	Utah
Where We Sleep: Homeless Cost Study	75% Cost Reduction	(Flaming, 2009)	Los Angeles
Denver Housing First Cost Analysis	$31,545	(Perlman, 2006)	Denver
Community Engagement Pilot Study	$42,075	(Moore, 2006)	National
Inebriate Program	73,352	(Dunford, 2016)	San Diego

Cost of Mental Health - Post Shelter Services – Statistics:

Source	Cost Savings per Person	Final results	Study
Jail Psychiatric Hospitals	$40,500	Dumping Patients	(Strickland, 2000, p. 1)
Hospital Psychiatric Wards	$22,500	Chronic Homelessness	(Poulin, 2010, pp. 1093-1098)
Sexual Assault	Exposure to HIV, Schizophrenia Cost	Victimization	(SAMHSA, 2013)
Fatal Accidents	Impaired cognitive thinking increases SSI & SSDI	Increased insurance costs	(Vella, 2014)
Unmet Psychiatric Needs	Costs 2.5% of GNP	Worsening mental illness Increased increases health care costs	(Andrews, 2001, p. 26)

The actual price of homelessness is diminished when you just count the chronically homeless, as reported in the study named "The Cost of Homelessness a Perspective from the United States" (Culhane D. , 2008, pp. 97-114). If on a given night approximately 407,966 individuals are homeless across the United States, and live in transitional housing, or shelters, according to a HUD survey, than at a $40,000 average cost per homeless the total United States yearly cost is about $14,500,000,000 yearly in housing alone (Moorhead, 2012).

Additional costs of homelessness are listed in the associated tables, and include children who develop poorer cognitive skills as a result of homelessness factors (Danseco, Evangeline, 1991, pp. 1139-1148).

Secondary Research

Psychosocial Value of Space:

Biologist Stephen Boyden's study shows that human well-being is optimized by the building architectures, and the social support the living setting offers (Heerwagen, 2008). Homeless shelters that are modest in size create caged environments that lead to increased repetitive motions, and pacing from the inhabitants which in turn lead to neurotic behaviors (Heerwagen, 2008) . This research reinforces the benefits that will be gained from combining a Vertical Farm design with a homeless housing project. The following benefits are also derived from Boyden study (Boyden, 1971), and identifies well-being needs that the vertical housing design combo will address (Boyden, 1971):

- Connection to Natural Environment.

- Meaningful Sensory Variability.

- Capability to hold and maintain personal comfort.

- Regular and enjoyable exercise.

- Psychological restoration.

- Chance for social equity, respect, and self-worth.

The Boyden study serves as critical support that shows comfortable indoor living spaces are critical in improving inhabitant's emotional outcomes (Thigpen, 2011, p. 15).

Green Collar Education and training:

The Sustainable Prison Project research outcomes showed an immediate positive impact from participants who engaged in the growing, and cultivating tasks associated with the program's agricultural training activities (Thigpen, 2011, p. 15). Participants who assisted in gardening experienced an improved connection to living beings, and increased sense of well–being that cannot be replicated through cognitive behavioral therapy (Thigpen, 2011, p. 15).

Job Crafting:

The Vertical farm project will include a job crafting component which allows trainees to design their own chores and work-boundaries. Employees who are allowed to change chores and social components of their respective jobs experienced an increased feeling of well-being, and self-worth and meaning (Wrzesniewski, 2013). Just as importantly, the job crafting component of this project adds to the positive organizational psychology of this program, which in turn elongates the project's lifespan, efficiency, and further protects the permanent housing status of housing participants (Wrzesniewski, 2013) .

2014 ANOUCEMENT - $100 MILLION COMMITMENT IN WORKFORCE FEDERAL GRANTS/FUNDING TO EMPLOYERS FOR JOB TRAINING (Labor, 2014)

TRAINED WORKERFORCE INNOVATION GRANTS OF 2014 - CONVERTING HOMELESS INTO LOYAL WORKFORCE (Whitehouse Staff, 2014)				
CITY	**AWARD AMOUNTS 2013**	**AWARD YEAR**	**Partnership Sector**	**Total Awarded to Partner, Employer**
New York	$12 Million	2013	City	Up to 80% of Salary Subsidized
Massachusetts	$11.5 million	2013	State Office of Labor Development	Upwards to 80% of Salary Subsidized
Los Angeles	$12 million	2013	City	Up to 80% of Salary Subsidized
Virginia	$11.9 million	2013	Community College System	Up to 80% of Salary Subsidized
Illinois	$11 million	2013	The Department of Commerce	Upwards to 80% of Salary Subsidized

The Importance of Centrality:

According to the literature, a trusting relationship with an all-in-one service provider is a critical service element for the success of any intervention for homeless families, and at-risk youth (Howe, 1998), (Barker, 2010). Thus, a provider who sets up a relationship built around trust with the homeless folks, and offers centralized addendum solutions such as training, psycho-social service, and etc. will more than likely achieve a successful homeless intervention program (Howe, 1998). Literature by Howe supports the importance of the role played by trust in improved social citizenship.

Community Reinforcement Approach (CRA)

The CRA literature produced is evidence that appropriate, and coordinated resources such as therapeutic interventions tend to be the most successful. (Selznick, 2008) (Meyers/Smith, 1995)

Monetized/Vocational Rehabilitation, Supported Program

Vocational Rehabilitation work-supported employment programs are evidence based interventions that motivate participants and result in positive outcomes (Staff, 2014). Additionally, vocational rehabilitation programs set-up for this program will improve client's mental health, increase rapid job findings, improve employment competitiveness, and benefit counseling (Hashemi, 2014). Conclusions of many of the studies utilized show that ongoing case management is vital for maximizing outcomes because ongoing participant examination increases outcome efficiencies, which in turn increases motivation of participants (Barclay, 2009, pp. 459-64).

Mixed-Use Buildings

Studies support how large, mixed-use, communities result in positive outcomes as it relates to participant's social interactions, and improvements within their cognitive behavior (Bailey,

2008). The research shows the importance of careful property management of this type of corporate system, and how such effective management will improve outcomes of the homeless (Bailey, 2008). The research also re-enforces the need to integrate a training component into the mixed-use community (Bailey, 2008).

IMPLICATIONS OF THE RESEARCH

Scaled-Up Homeless Shelters		Vertical Farming	
Mixed-Use Corporate Environment (Bailey, 2008)	Housing Sustainment, Self-supported, Self-contained	**Crop Protection** (Wikipedia, 2015)	Building protects crops from droughts, typhoons, & hurricanes
Higher Security Systems (Bailey, 2008, p. 3)	Eradicate Victimization & Minimize Criminal Activity Via On-site Property Management	**Crops Consumed Faster** (Wikipedia, 2015)	Less Spoilage, higher net incomes revenues
Integrated & Centralized Services (Bailey, 2008, p. 3)	Improved Wellness via family and community relationships	**Abandoned Properties** (Banerjee, 2013)	Productive usage
Community & Family Solidification (Bailey, 2008, p. 3)	Social Integration	**Limited Vehicular Transport** (Rose, 2015) (Jose, 2015)	Workers live & work on-site
Mental Health	Improved Dignity	**Overall Wellness** (Jose, 2015)	Less environmental bacteria

More Implications of Scaling Up into a Vertical Farm:

A 30 floor building with a foundation of a building equal to about 5 acres would generate a yearly crop equivalent to 2400 acres of traditional agriculture (Despommier, 2014). Architect Ken Yeang says that plant life should be cultivated within **mixed-used skyscrapers** in order to achieve maximum climate control (Wikipedia Staff, 2015).

Year round farming inside of a climate controlled building will multiply the productivity of the farmed surface by a factor of 4 to 6, and some research shows the factor may be as high as 30 (Wikipedia Staff, 2015).

Improved Wages and Employment from Housing Participants

This research indicates that housing participants will have improved job retention outcomes as a result of the implementation of best practices within this project (Culhane, 2010). Studies show that stable housing positively affects job retention (Culhane, 2010). Therefore, the net increase in wages for one inhabitant after one year of vertical housing support would increase to a minimum of $30,000 per year (Payscale Staff, 2015), and as per this project's requirement to supplement inhabitant's living wages.

RESEARCH PROBLEMS:

The price per square foot in large United States urban areas create problems for justifying Vertical Farming projects. But, that problem is overcome when the yearly cost allocated for homelessness is replaced with the minimal costs to implement Vertical housing farms.

Another problem could occur when substituting floor-space in buildings for competing urban land space. If this problem is too substantial for any given urban area than the project should be relocated to nearby USDA designated areas.

RESEARCH LIMITATIONS:

Vertical Farming is a relatively new idea, and therefore the research is limited in data. This can be overcome by approaching the project in a real estate listing model, and by taking research from comparable projects, and combining outcome similarities.

CHAPTER 2:

The purpose of this section of the research is to identify the most effective housing building selections, homeless service interventions, and strategies that provide quality housing in order to sustain families for the long-term. This is a different character model of a homeless shelter, in that this for-profit corporation will primarily serve as a real estate company that sells, and buys houses as just one of multiple ways to sustain the vertical-homeless housing project. In addition, our vertical farm project will generate 3,500 tons of fruit per year, along with 140 tons of catfish and tilapia as yearly sources of revenue (Banerjee, 2013, p. 4). Rain water and solar energy will be built into the development to cut critical operating costs. Also, the vertical farm component will feed housing residents, and generate enormous amounts of self-sustaining revenue for the project (Banerjee, 2013, p. 4).

Additionally, the project's parent corporation will work in collaboration with its ancillary non-profit housing corporations, in order to buy future investment housing at extreme discount rates, while generating massive income from farming and real estate activities, and simultaneously training and employing qualified homeless participants, thereby making a more sustainable homeless housing program, in conjunction with added financial support from government authorities.

Our end is to create a homeless shelter model that supports itself through good and bad times by collaborating with its corporate and government partners, and utilizing the participant corporation's subsidiary 501 (c) (3) funding capabilities. This extremely powerful and sustainable, corporate activist model, will continually pump revenue of 50 percent revenue generated from the parent for-profit corporation back into the farming and housing project.

Additional revenue funding will be ascertained from government financing sources, and workforce training subsidies.

Homeless Inhabitants will be integrated into communal and work areas of the 28 different, revenue generating, agricultural field invention designed to run off of solar energy, and wind (Kain, 2014). Vertical-housing farm project trainees will participate in a job crafting component which will allow them the freedom to design their own jobs, and work boundaries. Studies show that employees who are allowed to manage and change their task develop a feeling of increased well-being, and meaning (Wrzesniewski, 2013, p. 1).

Literature Review:

The research literature continually points to the design of the shelter as a deterrent to the homeless participants benefiting at the highest levels from the service due to the fact they do not feel the traditional shelter design does not meet their needs for dignity and self-worth. The literature points out that shelters are viewed as just another dominant, uncaring, institution designed to defeat self-worth, and built for purposes of social control (Sylvia Novac, 1996, p. vi).

An effective, sustainable, homeless design must combine the following two key design components (Thigpen, 2011):

- Buildings

- Individuals

In other words housing program planners must understand, and apply best practice variables to the project design based on how each of the above variables affects each design component (Thigpen, 2011).

Important parts of the secondary research utilized came from prison detention centers that learned the hard way regarding the negative effects inadequate humane building environments have on safety, health and well-being of the inmates, and the staff that supports their program (Thigpen, 2011).

Traditional homeless funding programs are proven by this research to cost the government more money over the long run due to their ineffectiveness, and costliness (Assembly, 2003, p. 6). Policy makers are moving toward a new approach to funding more effective social programs (Houstoun, 2015). Pay-for-success are now being used to fund social programs by investing in highly promising interventions as they succeed (Houstoun, 2015). Investments in a homeless model such as the one supported by this secondary research offer homeless participants affordable housing opportunities that result in a sense of dignity, and self-worth, as supported by the secondary research. This approach will lessen the cost burdens that chronic homelessness places on mental health services, prisons, hospitals, and improve the overall social sense of justice.

This housing program is structured to combine treatment for mental health, monetary aid, and case management. The literature illustrates that when homeless programs are infused into a singular community, and they combine case management with a myriad of additional services, than homeless outcomes improve exponentially (Rosenheck, 2003). The literature utilized concludes better outcomes for homeless when they are directly placed into supportive housing versus placed in a singular, multistage-models, at differing locations (Kasprow, 2009).

Conclusion of Literature Review – The research studies accessed for this project indicated several conclusions. The following are the key conclusions to be emphasized, and are derived from secondary research (Pable, p. 3):

- Limited living space results in a too crowded environment prompts teenage inhabitants to leave the safety of the shelter campus in search of breathing room.

- Child behavior and mental growth, improve when their bedrooms have additional size personal control features (Pable, p. 3).

- Families in smaller and altered living spaces showed negative territorial behavior such as space claiming and boundary setting (Pable, p. 3).

For every 10 years of Congressional "inaction" @ $40,000 per homeless the United States taxpayer spends upwards of $140 billion dollars for every 407,000 homeless (Moorhead, 2012) (Pauette, 2011).

SOLUTION:

The United States has over 100 tent cities, and the homeless cities are due to the homeless' need for safety, and community, according to an article by R. Kaufman. (Kaufman, 2014)

(Kaufman, 2014)

1st Floor: Intake, Housing Placement; & Social Services

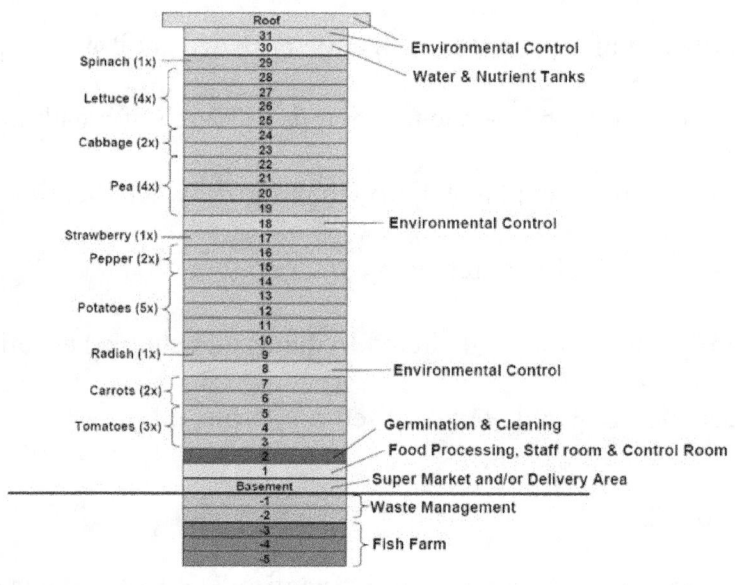

(Banerjee, 2013)

Interior/Surrounding Development Ideas

(ARUP Staff, 2013/2014)

(Kain, 2014)

Chapter 3:

Thesis Statement

The Thomas Charles Bass Realty homeless project provides an all-in-one services solution that includes housing services, and offers early childhood interventions, long-term housing, multi-faceted, and centralized psycho-social services. Our research report demonstrates how this cost effective, self-sustained, housing project inside of a vertical farming project will create more sustainable, and efficient homeless housing outcomes (Pinderhughes, 2011, p. 11).

Thomas Charles Bass, M.S.M.

MASTERS OF SCIENCE IN REAL ESTATE MANAGEMENT

"A California Real Estate Broker since 1992"

https://www.linkedin.com/in/thomasbassrealty

(800) 604-8436

Stakeholder Summary

Stakeholder pressure is increasing to find alternative homeless interventions, as a result of the mounting financial costs associated with accelerating national homelessness (Henwood, 2014, p. 4). The following are a list of key stakeholders who will participate:

- Residents of the Homeless Shelter

- Federal

- State

- City

- County

- Civic Groups

- Political Groups

- Corporate Stakeholders

- Special Interest Groups

- Study Groups

- Health Care Providers

Housing First, and similar service providers, have been critical in introducing new homeless program structures to stakeholders who are somewhat reluctant to change homeless logic structures (Henwood, 2014, p. 4). Establishing a partnership and incorporating Housing First into this formula could contribute to a faster assimilation as a key variable to stakeholders accepting this transformative change in the homeless shelter design (Scott, 2000, p. 3).

Project Approach and Methodology

Methodology is defined as a system of methods utilized in a particular area of study, as found at dictionary.com (Methodology, 2010). In ranking the most effect homeless system of methods, a cause-effect relationship criteria was used in parliamentary procedure to find outcomes or effects that led this research to determine the most desired outcomes for the homeless. The methods employed had three criteria that were used as the causal steps, which must occur within an effective housing program prior to the outcomes desired (Williams, 2003). The rubric used to determine which housing method was the most effective utilized the following steps (Williams, 2003):

- The cause must occur prior to the effect

- When the cause occurs the effect must "always" take place

- No other factors exist in the equation that can account for the effect.

Professor Dennis P Culhane says the greatest challenge is getting the people to overcome the stigma of the "shelter." (Lazo, WSJ, 2015) Therefore, the homeless center must scale up into a more dignified model to meet the "overall" emotional and service demand of the homeless in order to decrease the billions in cost associated with homelessness, as further inferred from Culhane's comments (Lazo, WSJ, 2015).

Hence, the intake center for this service should be on the bottom floor of this, new-aged, Vertical Farming, building, so as to immediately welcome and prepare housing participants for their new home. The intake center should offer easy access to immediate food, showers, clothing, social services, and immediate processing into permanent housing within the Vertical Farming, building, as further discussed by Lasso in the Wall Street Journal (Lazo, WSJ, 2015).

The all-inclusive intake center accepts potential inhabitants, their pets, stores belonging, and assigns permanent housing within the farming sectors (Lazo, WSJ, 2015). All participants must complete mental health assessments, as a mandatory part of the placement. Mental health workers identify the most severely mentally ill, and place them on high security floors 20-30. In these highly secured areas, participants who have more severe mental health needs are treated by mental health providers daily along with providing permanent housing and food. Once they are vetted and are mentally healthy enough they can work with their case manager to develop strategies for transition into the overall vertical farm community.

Less severe mentally ill participants, as determine by case managers, are sent directly to designated floors where training modules, and housing are assigned. Ongoing mental health groups and individual treatments are a requirement after intake.

But, permanent housing is not contingent upon employment. The basic requirement for permanent housing is compliant citizenry within the community. Family members are encouraged to visit ill family members, and family leave is granted as needed to enhance visitations.

Formerly homeless personnel are converted into productive employees who will trust the employer more, thereby cutting down on employee turnover. Managing the cost of employee turnover is key to keeping a stable workforce, and creates a more sustainable business model. The secondary research section solidifies this fact, and supports the significant cost savings to employers due to lower turnover (Boushey, 2012).

RESEARCH DESIGN

The qualitative and quantitative methods, likewise known as the mixed method, were both utilized for this research project. The qualitative method revealed the principal reason that the homeless stay away from many shelters is because "how they feel" regarding the stigma of homelessness (Guittar, 2014). This opinion is the main reason the city of Seattle, San Francisco, and Housing First have recently re-allocated their homeless housing expenditures by implementing scaled-up housing centers, and as evidenced by this statement by Ms. Rosen of Housing First: "Not only do you have support services on site, we build beautiful buildings and beautiful apartments," said Ms. Rosen. "You bring somebody inside, and you help restore their dignity. The support services that we offer help folks decrease their reliance on drugs. If they have mental health issues, they see a psychiatrist. And oftentimes their behavior is changed." (Moorhead, PolitiFact.com, 2012)

The quantitative method was applied primarily to support the most effective program service methodologies the homeless need. The most cost-effective methodology revealed through these quantitative studies combined the intake center with admission to permanent housing, and most importantly included mental health, and social services in the same fix, as further validated by a San Francisco city official in the Wall Street Journal (Lazo, WSJ, 2015).

Ethical Considerations

Shelter workers must follow the herein guidelines established to maintain professional boundaries between the homeless clients and shelter workers. The shelter's corporate group will maintain fair treatment, and good conduct at all times. Any deviation from these basic company policy standards will result in immediate termination.

Fair client treatment within this organization's context means acting in a caring, honest, and humane manner whenever interacting with shelter guests. The following are examples of this organization's daily ethical requirements:

- Maintain sanitary conditions
- Create and sustain a safe shelter
- Provide beyond adequate food provisions
- Client right to privacy
- Client right to self-determination
- Client right to dignity
- Client right to self-respect
- Staff cannot take gifts from guests
- No special favors for any customers
- No sexual relationships between customers and workers

These professional, ethics guidelines are basic to the code of ethics this center operates by. These principles along with fundamental morality principles will be observed at all times when operating within this system.

Chapter 4:

IMPLEMENTATION PLAN

Family Interventions

   A) Family Medical Assessments

   B) Housing Installation

   C) Stipend Orientation

   D) Group Therapy

   E) Individual Therapy

   F) Substance Abuse Management

   1) Therapy

   2) Monthly Blood Testing

   G) Parenting Skills

   H) Conflict Resolution

   I) Medication

   J) Job Training – Classroom

   K) On-the-Job-Training Module

   L) Workforce Collaborations

   M) Physical Exercise Intervention

   N) Housing Community Group Activities

Youth Interventions

   A) Early Intervention Programs

   B) Daily Nutritional Interventions

   C) Group Therapy

   D) Individual Therapy

   E) Life Skills

   F) Conflict Resolution

   G) Diagnostic Testing for Skill Interventions

H) Skills, Installation

I) After-School Tutoring (All Grade Levels)

J) Vocational Intervention

K) Physical Exercise Intervention

L) School-based Partnership/Intervention

M) Substance Abuse Management

    1) Therapy

    2) Monthly Blood Testing

VI. Homeless Service Delivery Protocol

Care Continuity

Customer Relationship Management

Case Management

EVALUATION PLAN

Return on investment is the best real world financial metric used to evaluate investment programs (Lynch, 2006, p. 55). Large investments, like the one needed for projects of this nature, require an assessment of the financial benefits accrued over time by investors (Lynch, 2006, p. 56)

One of the more standard measures of performance for property managers of this type of investment center is calculating the return on investment (ROI) (Staff, 2015). Investors tend to employ different ratios to determine if an investment is right for their purposes (Henderson, 2008).

The cash return for the taxpayer for a project of this magnitude would be astronomical. The following return on investment shows a more efficient use of taxpayer funds would result, and increased taxpayer revenue caused by increased ROI could be redirected and utilized in schools, training, the environment, and space travel thus making this country more competitive in the world.

Cash on return measures the return on cash invested in an income producing property (Henderson, 2008). The following are just a few of the fiscal benefits municipalities, and other stakeholder's investments in this program would achieve:

TYPICAL NET INCREASE IN WAGES IN ONE YEAR FROM ENTERING VERTICAL SUPPORT HOUSING				
Average Farmworker yearly Income (Payscale Staff, 2015)	**Previous Yearly Income estimate**	**Gain in Wages**	**New Minimum Taxable Income for Cities, State, & Federal**	**400,000 Homeless @ 2.6 tax** (Walczak, 2015)
$30,000	$0	$30,000	$30,000*400,000	12 Bill * 2.6% = $312 mill ROI tax gains
TYPICAL NET INCREASE IN WAGES IN ONE YEAR FROM DECREASE AGENCY FOOD ASSISTANCE REQUIREMENTS				
New Yearly Agency Liability per Recipient	**Previous Yearly Liability**	**Gain**	**New Revenue Income for Agencies Not Required to Pay Food Stamps**	**400,000** (Pauette, 2011) **Homeless Yearly ROI**
$0	$300 * 12 = $3600	$3,600	$3,600 * 400,000	$1.4 Bill ROI

TYPICAL NET INCREASE IN WAGES IN ONE YEAR FROM DECREASE AGENCY GENERAL ASSISTANCE REQUIREMENTS				
New Yearly Agency Liability per Recipient	**Previous Yearly Liability**	**Gain**	**New Revenue Income for Agencies Not Required to Pay Food Stamps**	**400,000** (Pauette, 2011) **Homeless Yearly ROI**
$0	$200 * 12 = $2400	$2,400	$2,400*400,000	≈ $1 Bill ROI

BUDGET – PROJECTED REVENUES - COST

VERTICAL FARM BUILD COST FOR A 60 STORY BUILDING & 15 ACRES (3 CITY BLOCKS SQ.) (Loopnet Staff, 2015)				
CITY	COST PER SQ. FT. (LoopNet)	TOTAL BUILDING SQ. FT. NEEDED	RENEWABLE ENERGY COSTS	TOTAL CONSTRUCTION COST (Plus Administrative Fees)
New York	$990	13,068,000 60 Stories	$140 per Sq. Ft.	15 billion
Los Angeles	$320	13,068,000 60 Stories	$140 per Sq. Ft.	6 billion
San Francisco	$680	6,534,000 30 Stories	$140 per Sq. Ft.	5.2 billion
Detroit	$40	13,068,000 60 Stories	$140 per Sq. Ft.	2 billion
Personnel	n/a	13,068,000 60 Stories	n/a	2.5 million
Fish Food, Plant Seeds, Nutrients, Equipment	n/a	13,068,000 60 Stories	n/a	190.5 million

Financing Sources From the New Holland Apartments Supportive Housing Projects (Mills, 2005, p. 16)	
Illinois Housing Development Authority HOME Funds	$1,802,140
Enterprise Community Investment Tax Credit (9%)	$2,864,679
Federal Home Loan Bank of Chicago: Affordable Housing Program	$1,166,663
Enterprise Environment Grant	$46,000
Illinois Clean Energy Community Foundation Grant	$211,402
Illinois Donation of Tax Credits	$214,000

(For donation of building)	
City of Danville Community Development Block Grant	$135,000
Enterprise Community Investment Reserves (9%)	$177,150
Cross-point Human Services (developer)	$249,816
Equity of Escrow (9%)	$166,500
Deferred Developer Fee	$43,500
	Total $7,260,250

(Mills, 2005, p. 16)

Conclusion

A critical aspect to a positive outcome for a pilot-project of this caliber is selecting a, licensed, real estate property management team with expert knowledge of the project and program. This firm has been a licensed real estate broker in the State of California since 1992.

In most states property management of this significant sized project must be managed by a licensed real estate broker (HG Staff, 2015). Performing certain real estate activities without a license could be a violation of the law, therefore a licensed, knowledgeable, real estate broker is vital to the legal operations of such a large size project of this magnitude which requires expert property management capabilities (HG Staff, 2015).

Also, a licensed property manager has been vetted by state background checks, and thoroughly educated in the practices of preparing demand letters, advising stakeholders and boards of directions regarding the state of laws relating to housing, farming, and return on investment issues (HG Staff, 2015). Brokers possess an excellent working knowledge, and information base related to market center factors which determine a farm's capacity to produce (Staff N. , 2014).

The United States Housing rental market pricing has skyrocketed, leaving hardworking families living in their cars, and shelters. Many a homeless family are found nesting in vacant and abandoned houses by realtors showing homes to clients.

The United States Census Bureau reported that 16% of the entire population live in poverty, and as of 2011, 16.7 million children live in food insecure households (Wikipedia Staff, 2015). This humongous number of food insecure households was a 35% increase from 2007 (Wikipedia Staff, 2015). Therefore, at a minimum of a 35% increase from 2011, or approximately 21 million children are anticipated living food insecure households this year.

Homelessness, often a result of an economic crisis which is exacerbated by crisis policy management. The government tends to design policy strategies to deal with sudden and significant events rather than developing long-term sustainable strategies (Rouse, 2013). The results of this poor policy management leads to job loss, mental illness, and increasingly more dangerous criminal activities (Staff, 2014).

As earlier discussed, in this report, the minimal cost of homelessness per person is approximately $40,000 (Pauette, 2011). This cost climbs astronomically when you take into account the $87 billion cost of incarcerating, and treating the mentally ill who often eventually become homeless (Torrey, 2010, p. 6). The research supports that transporting homeless people in emergency psychiatric emergency hospitals is a more costly, and less effective methodology than providing the homeless with a stable home, inclusive of supportive services (Kanter, 2010, pp. 1540-4560).

HUD data shows almost 400,000 can be homeless on a nightly basis (Pauette, 2011). And that number of homeless has grown at an extraordinary exponential rate yearly as a result of poor policy planning, and this report represents an evidence based solution to counter this disturbing trend. Investing in effective solutions of this caliber should be the **"number one"** priority of a great country like the United States.

Therefore, Congress' top legislative priorities should incorporate the following policies that will prevent homelessness, and alleviate hunger. Those policies should include, but not be limited to the following policy points (National Alliance to End Homelessness, 2015):

- Stop funding $39 billion dollars yearly for ineffective federal prisons (Henrichson, 2012).

- Stop spending $27 billion per year on failed drug policies (Staff, Get The Facts , 2014).

- Stop inefficiently paying billions nationally in emergency room hospital cost.

- Fund Vertical Farming "for-profit" projects that incorporate housing assistance.

- Repeal Dodd-Frank in order to strengthen small business job opportunities.

- Increase affordable housing policies at the national level to spur real estate.

- Lighten-up on underwriting overlays at the national level so that families can more easily afford to own a home.

- Decriminalize feeding the homeless at a national level (Levintova, 2014)

Corporate social responsibility is defined as a management concept whereby companies integrate social, along with environment concerns, into their daily business operations, and often it blurs the boundaries between government social services, and private industries (Silver, 2015, p. 3) (Ruggie, 2015). Vertical farming only becomes a viable revenue boosting concept when the sustainability garnered from its revenue, replaces a social plague like homelessness, and is nestled safely inside of a for-profit corporation, operating under a relatively regulation free government environment (Porter, 2006).

The United States is great because it was built by hard-working people with big ideas. This is a big idea that is supported by the data, **now it is time** to put the policies in place that will allow hard-workers to fulfill the prophecy of this great country. And, since policy makers cannot agree on simple policies that would improve social justice nowadays than "we the people" are now fully engaged, and herein operate as corporate activist partners with this plan.

Therefore, we the people support this corporate socially activist plan. We require that 50% of the e-book's net taxable income, and 50% of future net taxable income generated by the project to be used to sustain the research, and future growth of herein project. We concur that this project be incorporated in the future. And, in unison, with our continuing financial support of this project, **we speak into existence "The Bio-Climactic Vertical-Homeless-Housing-Farm."**

Works Cited

Allen, G. (2005). In G. Allen, Manufactured Home Merchandiser (pp. 53-54). RLD Group.

Andrews, G. (2001). In W. H. Report. France: WHO Library.

ARUP Staff. (2013/2014). Retrieved from Central Park, Shio: http://www.arup.com/Projects/Central_Park_Schio/Central_Park_Schio_Gallery.aspx

Assembly, C. D. (2003). Policy Guide on Homelessness (p. 6). Denver: https://www.planning.org/policy/guides/adopted/homelessness.htm.

Bailey, N. (2008, September). Retrieved from http://www.jrf.org.uk/sites/files/jrf/2295.pdf

Bailey, N. (2008, September). Retrieved from http://www.jrf.org.uk/sites/files/jrf/2295.pdf

Banerjee, C. (2013). Journal of Agricultural STudies, 46-47.

Banerjee, C. (2013). Journal of Agricultural Studies, 4.

Barclay, C. (2009). The contribution of IPS to recovery from serious mental illness. Retrieved from http://www.naric.com/?q=en/publications/volume-9-issue-2-psychiatric-disabilities-supported-employment

Boushey, H. (2012, November 16). There Are Significant Business Costs to Replacing Employees. Retrieved from Center for American Progress: https://www.americanprogress.org/issues/labor/report/2012/11/16/44464/there-are-significant-business-costs-to-replacing-employees/

Boyden, S. (1971). Biological Determinants of Optimal Health . International Biology Program. London: D.J.M. Vorster.

Carrier, S. (2015, March/April). Mother Jones - Room for Improvement. Retrieved from http://www.motherjones.com/politics/2015/02/housing-first-solution-to-homelessness-utah

Cilento, K. (2009, May 23rd). Dragonfly Vertical Farm concept by Vincent Callebaut. Retrieved from Arch Daily: http://www.archdaily.com/22969/dragonfly-vertical-farm-concept-by-vincent-callebaut/

Culhane. (2010). Wilder. Retrieved from http://www.wilder.org/Wilder-Research/Publications/Studies/Return%20on%20Investment%20in%20Supportive%20Housing%20in%20Minnesota/Return%20on%20Investment%20in%20Supportive%20Housing%20in%20Minnesota,%20Full%20Report.pdf

Culhane, D. (2002). Housing Policy Debate (pp. 107-62). Public Service Reductions Associated with Placement of Homeless Persons with Severe Mental Illness In Supportive Housing.

Culhane, D. (2008). European Journal of Homelessness, 97-114.

Danseco, Evangeline. (1991). Are There Different Types of Homeless Families. National Council on Family Relations, 46, 1139-1148.

Despommier, D. (2014). The Vertical Farm. Inhabitat.

Dunford, J. (2016, January). United States Interagency Council on Homelessness _ Impact Serial Inebriate Program. Retrieved from http://usich.gov/usich_resources/research/impact_of_the_san_diego_serial_inebriate_program/

English Online. (2015). Retrieved from Vertical Farming - Agriculture of the future: http://www.english-online.at/biology/vertical-farming/vertical-farming.htm

Flaming, D. (2009). United States Interagency Council on Homelesness. Retrieved from http://usich.gov/usich_resources/research/where_we_sleep_costs_when_homeless_and_housed_in_los_angeles/

Gudeman, S. (2015). Retrieved from http://www.green-buildings.com/articles/vertical-farm-what-is-the-cost-to-build/

Guittar, N. (2014). This is Where You are Supposed To Be: How Homeless Individuals Cope with Stigma. Sociological Spectrum.

Hashemi, H. (2014, September). Psychiatric Rehabilitation Journal. Retrieved from http://www.naric.com/?q=en/publications/volume-9-issue-2-psychiatric-disabilities-supported-employment

Heerwagen, J. (2008, May 23). Psychosocial Value of Space. Retrieved from Whole Building Design Guide: http://www.wbdg.org/resources/psychspace_value.php

Henderson, R. (2008). MRES Multi Real Estate Services, Inc. - How to Evaluate Investment Properrty. Retrieved from http://mres.com/how-to-evaluate-investment-property/

Hennessey, K. (2014, April 16). Los Angeles Times. Retrieved from http://www.latimes.com/business/la-fi-obama-job-training-20140417-story.html

Henrichson, C. (2012, January). The Price of Prisons. Retrieved from http://www.vera.org/sites/default/files/resources/downloads/Price_of_Prisons_updated_version_072512.pdf

Henwood, B. (2014). Journal of Community Psychology Practice, 8.

HG Staff. (2015). HG.org - Legal Resources. Retrieved from http://www.hg.org/property-management.html

Houstoun, F. (2015, April 1). Smart Management. Retrieved from Governing - The States and Localities: http://www.governing.com/columns/smart-mgmt/col-pay-for-success-social-impact-bonds-human-services.html

Howe, M. (1998). Cogprints. Retrieved from http://cogprints.org/656/

Jose, K. (2015). Retrieved from
    http://www.academia.edu/7655922/Vertical_Farming_Concepts_for_India

Kain, A. (2014, May 7th). The Dragonfly: A Giant Winged Vertical Farm for New York City.
    Retrieved from inhabitat - design will serve the world: http://inhabitat.com/dragonfly-
    urban-agriculture-concept-for-ny/

Kanter, A. (2010). Homeless, but not helpless: Legal Issues in the Care of Homeless People with
    Mental Illness. Journal of the Society for the Psychological Study of Social Issues;,
    Volume 45, 1540-4560.

Kasprow, R. O. (2009). Direct placement versus mutistage models of supported housing...
    Psychological Services, 190-201.

Kaufman, R. (2014). urbanful.org. Retrieved from Urbanful:
    https://urbanful.org/2015/02/11/how-do-the-homeless-fit-into-our-future/

Labor, U. D. (2014). Retrieved from https://www.dol.gov/dol/grants/FOA-ETA-15-02.pdf

Lazo, A. (2015, March 23). Retrieved from WSJ: http://www.wsj.com/articles/san-francisco-
    homeless-shelter-to-get-a-trial-run-1427144681

Lazo, A. (2015, March 23). WSJ. Retrieved from San Francisco Homeless Shelter to Get a Trial
    Run: http://www.wsj.com/articles/san-francisco-homeless-shelter-to-get-a-trial-run-
    1427144681

Levintova, H. (2014, Nov 13). Mother Jones. Retrieved from
    http://www.motherjones.com/politics/2014/11/90-year-old-florida-veteran-arrested-
    feeding-homeless-bans

Loopnet Staff. (2015). Loopnet Commercial. Retrieved from http://www.loopnet.com/Building-
    Properties-For-Lease/

Lynch, K. (2006). International Food and Agribusiness Management Review, 1.

Methodology. (2010). Retrieved from Dictionary.com:
    http://dictionary.reference.com/browse/methodology

Michael, C. (2014, February 13). Bright Agortech. Retrieved from
    https://www.brightagrotech.com/get-excited-about-vertical-farming/

Miller, A. (2013, December 23). Appfolio Property Manager. Retrieved from
    http://www.appfolio.com/blog/2013/12/should-you-renovate-your-rental-units-cost-
    versus-roi-is-only-one-consideration/

Mills, E. (2005, June 21). Evanmils - Costs and Benefits of Commissioning New and Existing
    Commercial Buildings. Retrieved from
    http://evanmills.lbl.gov/presentations/mills_cx_ucsc.pdf

Moore, T. (2006, June). United States Interagency - Estimated Cost Savings Following Enrollment in the... Retrieved from http://usich.gov/usich_resources/research/estimated_cost_savings_following_enrollment_in_the_community_engagement_pro/

Moorhead, M. (2012, March 12). Retrieved from PolitiFact.com: http://www.politifact.com/truth-o-meter/statements/2012/mar/12/shaun-donovan/hud-secretary-says-homeless-person-costs-taxpayers/

Moorhead, M. (2012, March 12th). Retrieved from http://www.politifact.com/truth-o-meter/statements/2012/mar/12/shaun-donovan/hud-secretary-says-homeless-person-costs-taxpayers/

National Alliance to End Homelessness. (2015). Retrieved from http://www.endhomelessness.org/pages/chronic_homelessness

Pable, J. (n.d.). Department of Interior Design, Florida State University (p. 3). Tallahassee: Research on the homeless population - the particular utility of case study methodology.

Pable, J. (n.d.). Retrieved from FSU : http://www.iiis.org/Pable.pdf

Pauette, K. (2011). Current Statistics on the Prevalence and Characteristics of People Experiencing Homelessness in the United States. Washington D.C.: SAMHSA.

Payscale Staff. (2015). Farm Worker Salary. Retrieved from http://www.payscale.com/research/US/Job=Farm_Worker/Hourly_Rate

Perlman, J. (2006, December 11). United States Interagency Council on Homelessness - Denver Housing First Collaboration Cost... Retrieved from http://usich.gov/usich_resources/research/denver_housing_first_collaborative_cost_benefit_analysis_and_program_outcom/

Pinderhughes, R. (2011, March). The Greening of Corrections - Creating a Sustainable System. Retrieved from fhi360 - The Science of Improving Lives: http://www.fhi360.org/sites/default/files/media/documents/20110426-Greening_of_Corrections508%20-FINAL_with%20disclaimer.pdf

Porter, M. (2006, December). Strategy and Society: The Link Between Competitive Advantage and Corporate Social Responsibility. Retrieved from https://hbr.org/2006/12/strategy-and-society-the-link-between-competitive-advantage-and-corporate-social-responsibility

Poulin. (2010). Mental Illness Policy Org. - Leaving Mentally Ill People on the Streets Costs.... Retrieved from http://mentalillnesspolicy.org/consequences/cost-homeless-mentally-ill.html

Rose, J. (2015, August). Retrieved from http://nashvillepublicradio.org/post/green-pie-sky-vertical-farming-rise-newark

Rosenheck, K. F.-M. (2003). Integrating health care and housing supports from federal agencies: An evaluation of the HUD-VA Supported Housing Program (HUD-VASH). Archives of General Psychiatry, 940-951.

Rouse, M. (2013, October). Retrieved from WhatIs.com: http://whatis.techtarget.com/definition/crisis-management

Ruggie, J. (2015). Harvard Kennedy School. Retrieved from http://www.hks.harvard.edu/centers/mrcbg/programs/csri

SAMHSA. (2013). SAMHSA - Behavioral Health Services for People Who Are Homeless. Retrieved from http://www.ncbi.nlm.nih.gov/books/NBK138716/

Scott, W. (2000). From professional dominance to managed care (p. 3). Chicago: University of Chicago Press.

Silver, A. (2015, March 26). Eat Up: Vertical Farming... Retrieved from http://www.scribd.com/doc/259967942/Eat-Up-Vertical-Farming-in-Sustainable-Cities#scribd

Staff. (2014). Retrieved from http://www.tgpdenver.org/homelessnessfaq

Staff. (2014). Get The Facts . Retrieved from http://www.drugwarfacts.org/cms/Economics#sthash.PGcLiDXz.dpbs

Staff. (2014). National Rehabilitation Information Center. Retrieved from Psychiatric Disabilities & Supported Employment: http://www.naric.com/?q=en/publications/volume-9-issue-2-psychiatric-disabilities-supported-employment

Staff. (2015). 2012 Books. Retrieved from http://2012books.lardbucket.org/books/accounting-for-managers/s15-04-using-return-on-investment-roi.html

Staff, N. (2014, July 30). National Association of Realtors. Retrieved from http://www.realtor.org/REALTORorg.nsf/pages/careers

Strickland, T. (2000, September). Retrieved from http://www.psychosocial.com/IJPR_13/Deinstitutionalization_Sheth.html

Sylvia Novac, P. (1996). No Room of Her Own: A Literature Review of Women and Homelessness (p. vi). Canadian Housing Information Centre.

Thigpen, M. (2011). The Greening of Corrections: Creating a Sustainable System (p. 15). Washington D.C.: National Institute of Corrections.

Torrey, F. (2010, May). Treatment Advocacy Center.

Vella, L. (2014). Policy Researchinc.org. Retrieved from http://ddp.policyresearchinc.org/wp-content/uploads/2014/07/Vella_Final_C3.pdf

Vertical farming. (2015, July 17). Retrieved from Wikipedia - The Free Encyclopedia: https://en.wikipedia.org/wiki/Vertical_farming

Walczak, J. (2015, April 15). Tax Foundation. Retrieved from http://taxfoundation.org/article/state-individual-income-tax-rates-and-brackets-2015

Whitehouse Staff. (2014). Retrieved from https://www.whitehouse.gov/the-press-office/2014/04/16/fact-sheet-american-job-training-investments-skills-and-jobs-build-stron

Wikipedia. (2015). Retrieved from Vertical Farming: https://en.wikipedia.org/wiki/Vertical_farming

Wikipedia Staff. (2015, August 2). Wikipedia. Retrieved from https://en.wikipedia.org/wiki/Vertical_farming

Williams, Y. (2003). Cause and Effect Relationship: Definition. Retrieved from Study.com: http://study.com/academy/lesson/cause-and-effect-relationship-definition-examples-quiz.html

Woods, L. (2007, December 30). Business Journal - - Property managers can help landlords increase revenue. Retrieved from http://www.bizjournals.com/portland/stories/2007/12/31/smallb2.html

Wrzesniewski, A. (2013). Advances in Positive Organizational Psychology, 282.

Zolfagharifard, E. (2014, January 27). Retrieved from DailyMail.com: http://www.dailymail.co.uk/sciencetech/article-2546585/Fields-orchards-SKY-Highrise-urban-farms-future-agriculture-claims-architect.html

---

Alaimo, Katherine; Olson, Christine (2001). Food Insufficiency and American School-Aged Children's

Banerjee, Chirantan (2015). Up, Up and Away! The Economics of Vertical Farming: Journal of Agricultural Studies; *Volume 2; 2166-0379.*

Barer, Justin; Humphries, Peter; McArthur, Moran; Thompson, Lorraine (2014) Literature. Effective Interventions for working with young people who are homeless; Australian Government; Department of Families, Community Services and Indigenous affairs;

improving the lives of Australians; Government Document;

https://www.dss.gov.au/sites/default/files/documents/06_2012/literature_review.pdf

Bonner, Sarah; Sprinkle; Geoffrey (2002). The Effects of Monetary Incentives on Effort…

Boushey, Heather; Glynn (2008). There are Significant Business Costs to Replace Employees:

Brody, G. H., Flor, D. L., & Gibson, N. M. (1999). Linking maternal efficacy beliefs, developmental goals, parenting practices and child competence in rural single-parent; African American families; *Child Development, 70*, 1197-1208; National Library of Medicine, National Institutes of Health; Journal;

http://www.ncbi.nlm.nih.gov/pubmed/10546340

Carpenter, Jeffrey; Myers, Caitlin (2010) Why volunteer? Evidence on the role of altruism, Image, and incentives; Journal of Public Economics, Volume 94; 911-920; Journal;

http://www.sciencedirect.com/science/article/pii/S0047272710000939

Carrier, Scott (2015), Room for Improvement; cleans up cities. Give the homeless a place to live, and Save money too. The shockingly simple, surprisingly cost-effective solution that won over a bunch of conservatives in Utah."; Mother Jones Magazine; March/April, Edition; The article discovered that 85 percent of the nation's estimated 84,000 chronically homeless which consist of men, women, youth, and whole families utilize transitional shelters on a nightly basis costing the government $30,000 to $50,000 per person in housing and social service services per year. The research highlighted a program named the Pathway to Housing. This apartment homeless shelter offered social services to residents only if they felt needed social service assistance. After five years, residents were found to be successfully dwelling in the residence, and utilizing the social services.

Also, the city showed a reduction of approximately $16,000 in monetary expenditure per

homeless person as a result of the Pathway program.

Center for American Progress -

https://www.americanprogress.org/issues/labor/report/2012/11/16/44464/there-are-

significant-business-costs-to-replacing-employees/

Cincotta, Frank; Arise Springiield (2012). Why I love Arise; why we work for social justice One;

http://arisespringfield.org/tag/shelter/page/3/

Cognitive, Academic, and Psychosocial Development; The American Academy of Pediatrics,

Volume 108; 44-53; Journal: http://pediatrics.aappublications.org/content/108/1/44.short

Crary, David; Leff, Lisa; (2014). Number of Homeless Children in America

Culhane, Dennis (2008). The Cost of Homelessness: A Perspective from the United States:

Dictionary.com (2015). Methodology definition. Dictionary.com:

http://dictionary.reference.com/browse/methodology

European Journal of Homelessness; *Volume 45; 97-114.*

European Journal of Homelessness, Volume 2.8; 97-114; Journal

http://repository.upenn.edu/spp_papers/148/

Fantuzzo, J., McWayne, C., Perry, M. A., & Childs, S. (2004). Multiple dimensions of family

Involvement and their relations to behavioral and learning competencies for urban, low-

income children. *School Psychology Review, 33* (4), 467-480; Journal;

http://psycnet.apa.org/index.cfm?fa=search.displayRecord&uid=2005-00702-001

Founder of Homeless center empowers homeless children by teaching them photographic skills.

The children utilize the photography skills to tell their homeless story.

Hellegers, Desiree (2011). No Room of Her Own: Women's Stories of Homelessness, Life, Death, and Resistance; Palgrave Macmillan; 218 Book.

Howe, D.C. (1998); Participation, Citizenship and Trust in Children's Lives; Book https://books.google.com/books?id=RUkic_aYg6IC&pg=PA3&dq=Howe,+1998+trust+r ole&hl=en&sa=X&ei=AchrVdraNIXAggTx4oCICQ&ved=0CB4Q6AEwAA#v=onepage &q=Howe%2C%201998%20trust%20role&f=false

Hubbard, Jim (1996). Lives Turned Upside Down: Homeless Children in Their Own Words and

Julia; Bowne, Pete; Grady, Cheryl (2015). Neutral Responses to Monetary Incentives in

Kain, Alexandra, Inhabitat (2014).The Dragonfly: A Giant Winged Vertical Farm for New York

http://inhabitat.com/dragonfly-urban-agriculture-concept-for-ny/

Women share their experiences of homelessness. The book describes the horrors of drop-in centers, and unkempt shelters. The book brings home the importance of quality housing and the effects of negative housing shelters on the human mind.

Kim, Eric (2006). 103 Things I've learned About Street Photography; Digital Photography School, http://digital-photography-school.com/103-things-ive-learned-about-street-photography/

Kushel, Margot; Vittinghoff (2011). Factors Associated with the Health Care Utilization of Homeless Persons; American Medical Association, Volume 285; Journal - http://jama.jamanetwork.com/article.aspx?articleid=193438

Lazo, Alejandro (2015). San Francisco Homeless Shelter to Get a Trial Run: The Wall Street Journal: Online - http://www.wsj.com/articles/san-francisco-homeless-shelter-to-get-a-trial

McCarty, Dennis (1991). Alcoholism, drug abuse, and the homeless: American Psychologist

Volume 46; 1139-1148 Print. This report indicates that certain interventions controlling and rehabilitating the homeless regarding alcohol abuse contribute to the effectiveness of programs for the homeless.

Application for rezoning of 15 acre parcel to be changed from single-family to multiple family zoning classification. The complainants alleged that the zoning denial was racially biased and prejudiced.

Morris, Alex (2014), the Forsaken: A Rising Number of Homeless Gay Teens Are Being Cast Out by Religious Families; While life gets better for millions of gays, the number of homeless LGBT teens - many casts out by their religious families - quietly keeps growing"; Rolling Stone, Magazine: http://www.rollingstone.com/culture/features/the-forsaken-a-rising-number-of-homeless-gay-teens-are-being-cast-out-by-religious-families-20140903. A San Francisco State University study empirically shows that highly religious parents are significantly more likely to abandon their children in the streets for being gay. The article discusses some rationales why LGBT people make up 40 percent of the homeless.

Markee, Patrick (2014), Research Proves that Federal Housing Programs Work to

O'Connor, Hollie; MySA, (2011). Homeless children capture their dreams in photos; 'Pictures of Hope' program let youngsters focus on their future: http://www.mysanantonio.com/news/local_news/article/Capturing-their-dreams-2188569.php. Photographs; Simon & Shuster, 6th Ed. Book

Paul, Lauren (2015), Not in Newton's Back Yard; Well-heeled progressive champion liberal ideals,

Including housing the homeless. Just don't try it in their neighborhood; Boston Home;

Magazine: http://www.bostonmagazine.com/property/article/2015/02/24/affordable-

housing-in-newton/.

This magazine article talked about a city of Boston homeless shelter proposal, if

accepted, that would permit the developer to use $1.4 million of city controlled housing

funds toward a $3 million renovation project.

Reduce Family Homelessness; Coalition for the Homeless:

http://www.coalitionforthehomeless.org/wp-content/uploads/2014/03/BriefingPaper-

ResearchonHousingAssistanceforHomelessFamilies-2-12-2009.pdf

An academic research briefing paper that demonstrates the strength of federal funding

housing assistance plans. This public research covers a 20 year period and illustrates the

implications of increasing support for federal housing programs.

Study.com: http://study.com/academy/lesson/cause-and-effect-relationship- definition-examples-

quiz.html

Surges to All-Time High: Huffington Post. http://www.huffingtonpost.com/2014/11/17/child-

homelessless-us_n_6169994.html

Stivers, Laura (2011). Disrupting Homelessness: Alternative Christian Approaches; Publisher -

Fortress Press; 187 Book.

Offers solutions communities can create social movements to alleviate poverty and

homelessness. The writer explains why many current approaches to homelessness are

non-effective.

Tanvirul Islam (2011). File: Street Child, Srimangal Railway Station.jpg; Wikimedia Commons,

http://commons.wikimedia.org/wiki/File:Street_Child,_Srimangal_Railway_Station.jpg

Task Performance; theories, evidence, and a framework for research; Accounting, Organizations

and Society, Volume 27; Issues 4-5; 303-345; Journal -

http://www.sciencedirect.com/science/article/pii/S0361368201000526

U.S. Dept. of Housing and Urban Development (2010). Employment and Training

Administration. Workforce Innovation

Fund Grant Awards: European Journal of Homelessness,

http://www.doleta.gov/workforce_innovation/grant_awards.cfm

U.S. Dept. of Labor (2011). Employment and Training Administration. Workforce Innovation

Fund Grant Awards: European Journal of Homelessness,

http://www.doleta.gov/workforce_innovation/grant_awards.cfm

Village of Arlington Heights Et Al. V. Metropolitan Housing Development Corp. Et Al: 1976:

Justice Powell; Chicago Press, 19-77;

Wikipedia (2015). Vertical Farming - https://en.wikipedia.org/wiki/Vertical_farming

Williams, Yolanda (2012). Cause and Effect Relationship: Definition, Examples &

Younger and Older Adults; Brain Research, Volume 1612, 70-82; Spaniol, Julia; Bowne, Pete;

Grady, Cheryl